Trees That Twist

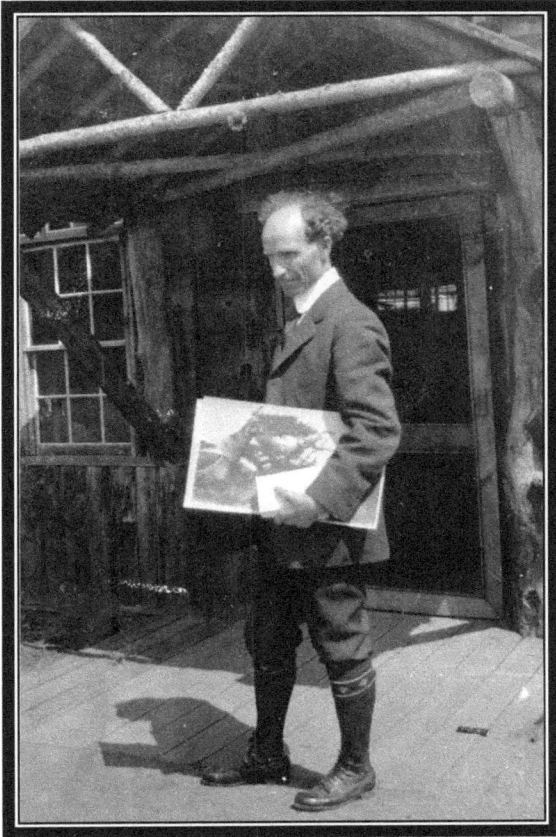

Enos A. Mills

Temporal Mechanical Press
Longs Peak, Colorado

Temporal Mechanical Press
a division of Enos Mills Cabin
6760 Hwy 7
Estes Park, CO 80517-6404
www.enosmills.com
info@enosmills.com

ISBN 978-1-928878-26-1

Title Page: Enos A. Mills, holding a copy of his photograph
of a twisted pine.

Back Cover: Enos A. Mills with his daughter, Enda, in a
twisted pine.

Introduction

This little book was published many years after Enos A. Mills' death by his widow Esther while she owned Long's Peak Inn. So many people have enjoyed it we have included a forestry speech which Enos gave that people have shown interest in, and some new photographs.

We hope you like this enhanced book and it will renew your respect and compassion of these magnificent spirits, our neighbors and friends.

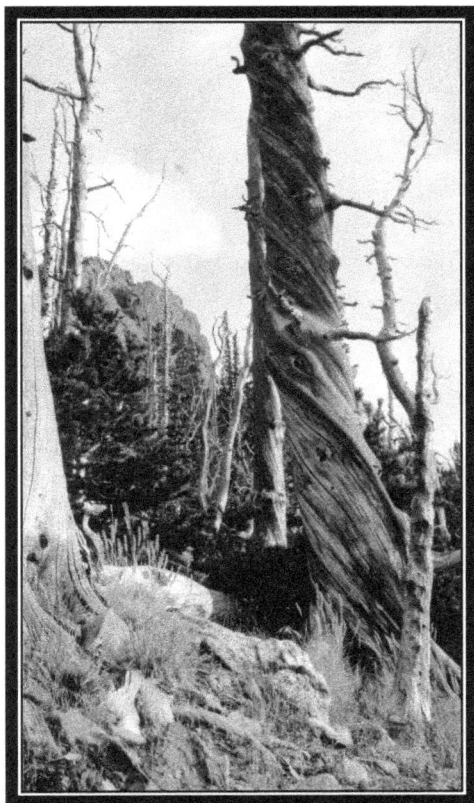

"Life is either a daring adventure or nothing. Security does not exist in nature, nor do the children of men as a whole experience it. Avoiding danger is no safer in the long run than exposure."
Helen Keller

Trees That Twist

During years of wandering through the forest I found twisted trees of many kinds and in many places, and I frequently asked myself, why do trees twist? The twisted grain of trees proved to be an interesting study in tree physiology.

One day I watched a bear start to cross a swift mountain stream on a slender, fallen tree that made a forty-foot bridge. The water was high, and it roared and surged as it entered the mouth of a deep canyon. The rocky banks rose twelve or fifteen feet above the water. Without the least hesitation the bear stepped upon the log and started to walk across. The log at once began to sag beneath his weight. Bears are good swimmers and I expected after two or three steps the log would break and I would see him in the swift current. But he only turned his head to watch the sticks that were floating swiftly along beneath him. As he approached the center of the log he struck playfully at one of the passing objects, although his paw did not reach within four feet of it. He caused the sagging log to bob up and down, and plainly enjoyed this movement. Without any mishap the bear proceeded across the log, lingered a moment on the opposite bank to listen to the roar of the water, and then went on up the mountainside. I wondered if he had ever crossed this log before and knew how tough it was. I went over to examine it.

This log was wet and toughened from a rain of the preceding night. But without any wetting it would have been a tough one. It proved to be a Douglas spruce, a tree whose wood is most sought for airplane timber, and which has for a century furnished the ship masts on the seven seas of the world. In places the bark was torn off, and as I examined it I discovered that the fibers of the wood were twisted round and round like the turns of a screw. I chopped deep into the tree and found that it was twisted nearly to the center. Then I walked across and cutting off the top of the tree and found there it was twisted through to the center. Returning to the other side of the stream I chopped off the stump with its upturned roots and found that the twist at the bottom had begun when the tree was about two inches in diameter. This spruce had grown close to the bank of the stream with its roots wedged in the vertical and horizontal cracks of the rocks.

In questioning numbers of people about twisted trees I was told that the wind was the chief cause of this deviation from the normally straight grain. This seemed plausible, but after finding pronounced twists in trees in sheltered coves and in the bottom of windless canyons, I abandoned the wind as even a possible factor.

Several lumberman told me that soft woods such as pines, firs and aspens twist from left to right, and the hard woods from right to left. And I was also repeatedly told that trees in the northern hemisphere twist only from left to right and that below the equator they twist from right to left. This I believed until one day I found contradictory evidence. I had photographed a dead pine of pronounced spiral growth, twisting from left to right. In developing the negative I discovered another pine standing a few feet from the twisted tree that had a twist in the reverse direction.

Limber pine and Douglas spruce are the two species, of the trees I have examined, which show the most extreme twists. I have also found specimens of twisted trees among the western yellow pine, the lodgepole pine, the aspen, elm, cottonwood and two or three species of oak. It may be that all species twist, though less intensely, whenever there is an influential combination of resistant conditions.

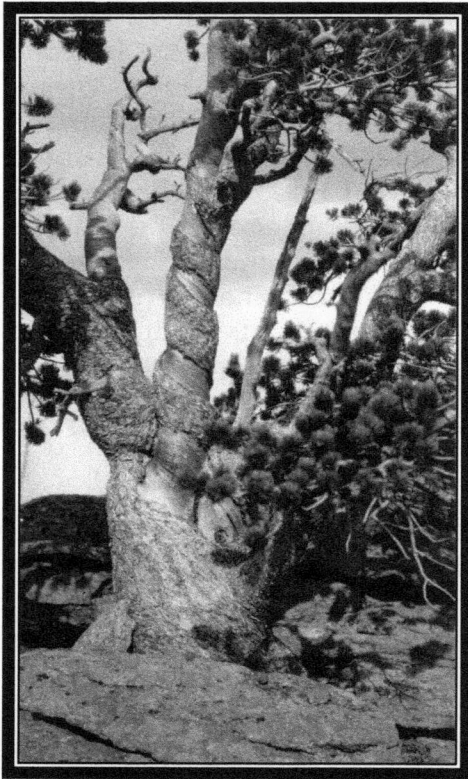

At Left: A twisted
Limber Pine.
Lightning tore the
spiral pathway in the
bark, following the line
of lease resistance.

A limber pine on the slope of Long's Peak,
killed in the forest fire of 1900.

One day I examined a dead pine in which the turns of the spiral were so close together that the log had the appearance of being wrapped round and round horizontally with a grayish-brown yarn of wood fibre. But in most trees the twist is more nearly that of a much elongated spiral spring, going around a tree of one foot diameter once in every two feet of rise. The spiral paint on a stick of candy illustrates the closeness of the turns in the twist of most twisted trees. The limbs of twisted trees appear to be twisted in the same direction and in like degree as the trunk.

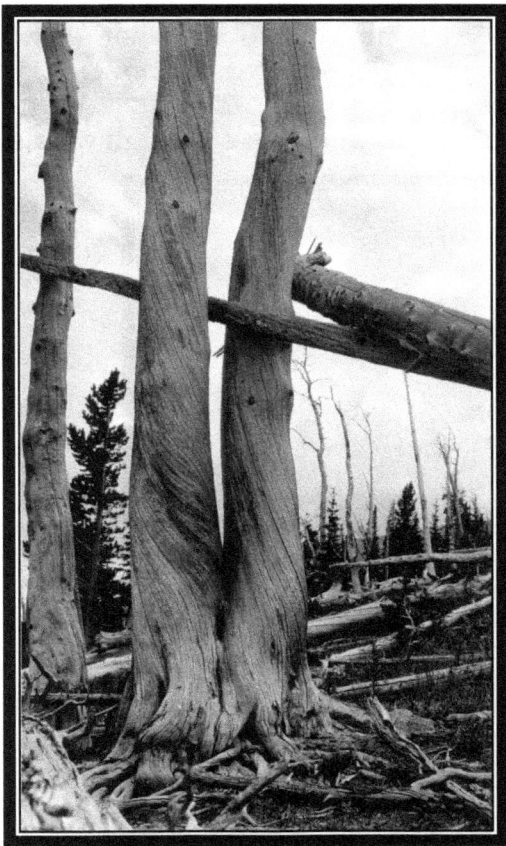

Such trees, if securely anchored, may stand for a generation or two after being killed by fire. They do not split easily, and resist the wind.

It is almost an impossibility to split these larger twisted trees. Limbs or knots run through the coiled spirals like so many iron bolts or reinforcing rods of steel. I have tried splitting short twisted sticks when they were wet. I had exercise but did not add to the pile of split wood. When they are frozen, however, they can be split.

Trees of this twisted type are more durable for outside timbers; and when sawed they often are more attractive for veneer or for polish than the woods of more straight grained trees.

The most closely twisted trees that I have seen are the limber pine of the Rocky Mountains from an altitude of 8,000 feet to Timberline, somewhat above 11,000 feet. And the most intensely twisted specimens of these were trees that grew in the most trying conditions at timberline—contending with high winds,

drouths, sudden changes of temperature and an excess of rocks in the soil.

The twist in a living tree rarely shows unless the bark has been removed from some cause. I have had the best opportunities to study twisted trees in fire-killed areas where the bark has been burned off, and at timberline and other wind-swept places where the bark has been sandblasted away, and in trees struck by lightning where the lightning has torn its spiral pathway in the bark.

Even lightning is given a spiral journey by the structure of a twisted tree. When lightning strikes a tree of ordinary straight grain it usually runs directly to earth, plowing a furrow, bark-deep, along the way. But when a twisted tree is struck the bolt generally follows the spiral fibres round and round the trunk to the ground. This course is evidently the line of least resistance. I have found a few of these twisted trees which were completely wrecked at the base by the lightning bolt. On entering the roots the electricity appears to have exploded, tearing the base to pieces and hurling roots and rocks in every direction. But the majority of lightning-struck trees, as well as straight-grained trees, are not seriously injured by the bolt.

An old prospector thought that dryness was the chief cause of this twisted grain, and took me to a dry, wind-swept ridge where most of the trees were twisted, as their naked trunks revealed. But we examined trees in a nearby wet place and removing the bark from a number of dead trees, found that they, too, were twisted. Both those in the dry and in the wet places were rooted in and upon almost unshattered rock.

Rock resistance appears to be the chief cause of twisted tree growth. I have a photograph of two limber pines which grew side by side in a violently wind-swept place. One of these is twisted and the other straight-grained. The straight-grained one is standing in deep

At Right: Photograph by
Esther Burnell Mills.

Below: An intensely
twisted limber pine at
timberline, sand-blasted
by the wind.

Above: A pine with a left-hand twist.

At Left: A pine with a right-hand twist.

soil and the other in shallow soil rooted in the cracks of a nearly solid rock.

In building a short stretch of mountain road, hundreds of small lodgepole pines were uprooted and thrown out. I examined many of these and found that the majority were more or less twisted. Of the sixty-four which I examined, fifty-seven had a right-hand twist.

Years later I was coming through a standing fire-killed forest of lodgepoles. The bark blistered by the fire had recently dropped off, revealing the texture of the wood. Many of these trees were twisted. I circled around this spot until I had counted sixty-four twisted trees, fifty-three of which showed a left-hand twist. Apparently there is an inherited tendency in trees to twist, which may locally be either to the right or to the left. A single lodgepole with hoarded seeds in a burned over area may have a thousand descendants within a stone's throw. If it is possible to transmit a tendency either to a right or a left twist, it would be easy to understand the distinctive character of these two lodgepole pine tracts.

I found seedlings no larger than a lead pencil that were twisted, but many trees, like the one the bear used for a bridge, did not begin to twist until they were two or more inches in diameter.

Apparently, the roots of a young tree work their way into the cracks and crevices in rocks where they find all the available soil and much moisture, and the best possible anchorage. After a few years' growth, the roots have expanded, completely filling the cracks and crevices, which in many cases are small. The growth continuing, the rocks exert a choking pressure upon the expanded roots. In some cases I found the roots not round but flattened, and spread out into what might be called a network of long, slender, thin fingers, two or three feet long, but at no place more than a half inch thick.

Just how this resistance of the rocks is related to the development of the tree roots, to the gathering of nourishment, water, and the food which it contains, plant physiology may sometime explain. But every twisted tree that I have examined grew in a rocky place where the roots were wedged among crevices in almost solid rock that could not be split wider apart. Twisted trees are picturesque and among the thousand and one interests out-of-doors.

Two side-by-side trees showing marked differences in how they twist.

This page: Elizabeth "Bessie" Burnell, guide at Long's Peak Inn and Enos' sister-in-law.

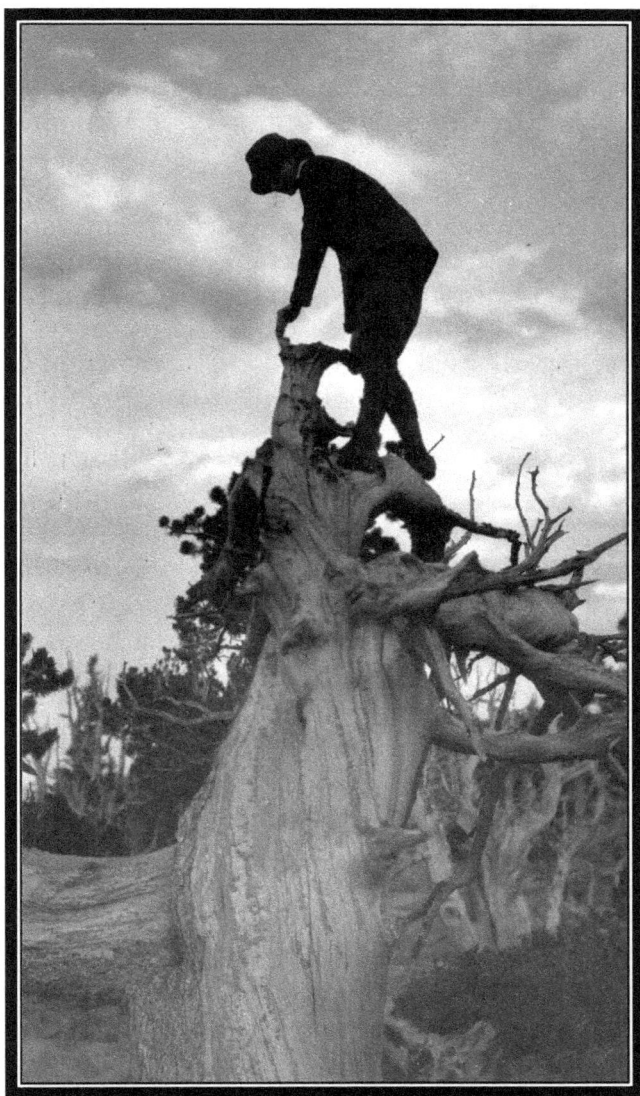

Esther Burnell Mills, guide at Long's Peak Inn, and Enos' wife.

The Forests of Colorado

Enos A. Mills address delivered by the request of the
Denver Chamber of Commerce, at a luncheon,
March 9, 1905

The present forest of Colorado is a picturesque remnant—a melancholy ruin of former grandeur. Christmas
trees are becoming scarce. Lack of forest sentiment has
made us "land skinners" of us all. The forests of Colorado are mostly evergreen, and are made up of beautiful
and useful trees which give use and beauty to the grand
mountain slopes on which they grow.

Do not the people of Colorado need a forest to
supply water to the farmer, timber for the miner, lumber
for the cottage, wood for the hearthstone, Christmas
trees for the children, climate for everybody and scenery
for us all? No country has ever grown poor by maintaining a forest. Spain has poverty and Syria has desolation,
but these conditions were not caused by forests. If
Colorado wants a forest for use, beauty and perpetua-
tion she is using the wrong remedy. Mr. Dooley says "It
makes no difference what kind of doctor ye have so
long as ye have a good nurse." The forests of Colorado
seem to have a very bad doctor, and no nurse at all.
Americans have looked upon the forests as something
in the way. If there are to be forests there must be forest
sentiment. Sentiment is formed in childhood.

Young Colorado sowed wild oats with the inherited
forest fortune. But it is good to know that this young
state is realizing with Mr. Dinkelspiel that wild oats
make poor breakfast food. In Colorado the pioneers
found about 36,000 square miles of forest. This amount
would put a mile-wide forest belt round the world. Of
this forest, the main belt, about 25,000 square miles has

been burned, and much of the remaining overlap of the belt has been used. Perhaps 5,000 square miles, or about one-seventh of the original, remain. But this remainder is not all good forest.

Much of it was too scattered to burn, or too scrubby for lumber. A great proportion of the regrowth on the burned woodland is of species and quality inferior to the original forest. Much aspen is growing where spruce grew and should grow. Probably one-fifth of the burned area has a close regrowth; some of the charcoal area has a straggling number of young trees, but upon about one-third of Colorado's woodland there is no growth worth considering.

Much of the treeless land could be reforested with but little expense. The soil of Colorado generously responds to all seed sowers. In many places in the state may be seen thousands of young trees growing upon the strip over which a snowslide went. The slide harrowed in the seed. If the reforesting is left to Nature she should be given a chance. Forest fires are now invading the regrowth. Forest fires make no distinction—they will alike devour the trees of a scientific forester, or those planted by Nature. The law which requires sawmill men to burn the slash seems a bad one. Fires from slash piles often kill nearby trees. These fires also consume much valuable mould and burn many seedlings. Left upon the ground, the slash would accumulate and conserve moisture, and form excellent protection for the many young trees which would grow up among it. Both fires and man seem determined to exterminate Colorado's little forester, the tiny, scolding and active Douglass squirrel. Each year a surprising number of trees grow from the cones planted by this little gray squirrel.

The beaver is another animal which is a friend of the irrigator and the forest. But steel stratagems and fires have almost exterminated the beaver. Formerly the beaver maintained thousands of dams and ponds along the upper courses of the streams. These dams and ponds checked floods, prevented erosion, delayed the run-off and did much to give a regular, constant flow to the stream. The beaver and the Douglass squirrel are well worthy of a chapter in that great book, "The Earth as Modified by Man."

Wherever forest rangers have been used their presence greatly tends to prevent forest fires. In Germany, where all land skinners are dead, they not only have forest rangers, but they also have watchmen stationed in towers or on promontories in the forest to watch for fires. These watchmen have fire-fighting tools at hand, and also a telephone with which to send in an alarm. Perhaps the greatest needs of Colorado's woodlands to-day are forest rangers and fire watchmen. Private as well as public wood-land should have fire protection.

Fires have made the Rocky Mountains still more rocky. Much tree seed scattered by Nature falls on stony ground. In many places fires burned everything to the solid rock. In other places the partial havoc of the fire was completed by wind or water. Fires and accompanying evils have literally skinned off acres of soil. Much of the Rockies is a hard granite which disintegrates slowly, and Nature has a tedious task of preparing soil. Fires not only mean a loss of forest, but also a loss of food upon which a new forest might be fed.

The cedars of Lebanon grew upon the moraine of a glacier, and the greater portion of Colorado's heavy forest is growing either upon moraines, or on the soil beds formed during the last glacial epoch. Most of the glacier lakes were long since filled with debris, and many of these old lake basins now form those beautiful flower meadows or the grassy vistas which are so delightfully embowered in our alpine forest.

The principal woodland zone lies between 7,000 and 11,000 feet altitude. Upon this in altitudinal distribution the different species of trees are to a great extent distributed upon the ground which most nearly meets their requirements for light, heat and moisture. However, trees, like people, have to struggle for existence, and many fail to obtain or to hold the place for which they are best fitted. The mountain slopes descend toward all points of the compass and have a variety of climate, but the distribution of trees over these slopes is quite regular. Trees have tongues, and should one acquainted with the language awaken in the woods he would, without other means, be able to tell the points of the compass, the altitude and even the season of the year. Aspen is the only species everywhere found over the tree zone, but in only a few districts does it rise to the stature of a tree, or mass its big trees into a forest. Local conditions here and there force timberline up and down the slope a little, but for the most part it is below 11,500 feet.

At timberline, Range pine, Arctic willow and spruce are found in stunted growths. Engelmann spruce has almost complete possession from just below timberline down to the 10,000 foot level, where White and Range pine and Alpine fir begin to intermingle. Descending a little lower Balsam fir and Red spruce are found, and just below 9,000 the upper ranks of the yellow pine. At 8,500 the Silver spruce flourishes along the streams, while below 8,000 to lower margin of the forest belt is the district on which Black pine, Pinion and Cedar are found. The Pinion, both in size and in scattered growth, reminds one of an orchard tree, It makes excellent fuel and bears nuts. There is an extensive forest of Cedar and Pinion on Mesa Verde, in the southwestern corner of the state. But this forest gives the ground very little shelter, the cedars being squatty and selfish, holding each at bay with long arms.

The distribution just given is in places broken into by local growths, and on the eastern slope of the front range there are extensive tracts of Lodge Pole Pine. Silver spruce and Balsam fir can be but little used for lumber, but they are Colorado's most beautiful trees. Both court the mountain brooks of the middle forest district and are the evergreen poems of the wild. They get into the heart like a hollyhock. As attractions for tourists they rival the grand peaks themselves. The Balsam fir is precisely artistic, is a slender pyramid of fern-like limbs, delightful in beauty and rich in aroma.

The Silver spruce is truly Colorado's most handsome tree. Her beauty of form, her fluffy silver-tipped robe, her garlands of rich brown cones make her the queen of Colorado's wild gardens. Red and Engelmann spruce, Yellow and White pine, are probably the best known and most used trees. Engelmann spruce embowers the streams and is the best water conserver in the forest. This tree prefers the cool slopes, and it wades the swamps and crowds the gulches. Seven thousand, or even more, crowd upon an acre, and thus form an excellent cold storage for winter's ice and snow, sometimes keeping these in store all summer. The splendid Engelmann spruce is bound to get into the literature and the hearts of those who live where irrigation means life.

I never see an Engelmann spruce without thinking of Henry Michelsen. It was the tree he loved. He appreciated its worth, and sadly he saw this noble tree go down in the relentless slaughter. Henry Michelson wore out his life in order to perpetuate the forest of Colorado. He did an enormous amount of valuable work, not only for the reserves of which he was superintendent, but incessantly labored for forest welfare, and thus for Colorado. Without pay, receiving only contempt for it, he has risen from his bed at midnight to make out the necessary papers which were the next day to compel indifferent county sheriffs to fight neglected forest fires. He was an unselfish man, a splendidly equipped forester, and he supported his intelligence and honesty with utmost daring. Henry Michelson gave his life for the forests of Colorado. The last time I saw him he said to me: "When you come back be sure and tell me all about the Engelmanns on the North fork of the Grand." When I came back he was dead. But he did not die in vain. Soon upon Colorado's thousand hills an evergreen forest will immortalize his memory.

Red spruce and Yellow pine meet on the hilltops of the middle forest zone. The spruce crowd the cool places, while the pines scatter over the dry, sunny slopes. Yellow pine has a stocky body and strong limbs. Its rich brown bark is almost asbestos, and is of a rich golden brown. Yellow pine has the most character of any tree on our hills. There are many fire-scarred veterans in its ranks, and when evergreens fight fire it is the last to fall. It supplies the pine knots for the fireplace, and makes good heavy lumber, but it cares little for the water supply either for itself or for anyone else. Red spruce does not fight fire very well, but old or young, it generally manages to hold the territory of its fathers against the most persistent invaders. It is not at all like its Yellow pine neighbor, but looks after the water supply carefully. It makes excellent mining timber and good lumber.

Rainmakers cannot be relied upon, so the next best thing to be done is to take care of all the water which Nature supplies. Considered for water supply alone, is not the forest of inestimable value? Such able men as Professor Carpenter and Mr. Michelson say the forests conserve the water. Perhaps the most valuable witness in all the world on this subject is John Muir. He says: "Drain off the water and the trees will remain, but cut off the trees and the streams will vanish."

Most all scientific foresters hold that forests will increase the water supply by checking evaporation. Increase the water supply of this state and you increase the number of cottages. The chief water loss is by evaporation. Colorado winds are extra dry, and their thirst is insatiable. They steal a heavy percentage of the entire water supply. The measurements of this and other governments show that more water comes from forested watersheds than from barren ones, and measurements in abundance show that streams flowing between forested banks rarely show even a small loss from evaporation, while unsheltered streams are constantly wasting on the air.

I believe that every one in Kansas and most every one in Colorado realizes that Professor Carpenter is thorough. A few years ago he brought out bulletin 55, "Forests and Snow," in which he placed a variety of evidence which shows the great use of the forest in saving water. Last May, on the day of the Poudre flood, I came down off the range, and on the South Poudre found snow from three to four feet deep in the woods just below timberline. When within about two miles of Chambers lake I came to a large burned area. The altitude at this place was about 9,800 feet, and all through the green woods the snow lay from twenty-two to twenty-eight inches deep. But when I stepped out into the burned area I kicked off my snow shoes and walked upon the earth. There was no snow at all.

Early last June, 1904, I explored the headwaters of the South Boulder. At between ten and eleven thousand feet, in the green timber along the Moffat road, the snow lay, on an average, five and one-half feet deep. A few miles south of the road there is an extensive area devastated by fires. In this area on a similar slope, and at the same altitude as where I found the snow by the road, there was but little snow to be found. There was an occasional drift in the shelter of a gulch, or in the lee of a crag, but for the most part the ground was absolutely bare. There are scores of similar instances not only from other places, but from other years, which emphatically demonstrate the enormous value of the forest to the water supply. Last season the government conducted a series of experiments concerning evaporation. From a water surface at Logan, Utah, more inches of water were evaporated within three months than fall upon Colorado each year. The daily evaporation at this place was .273 in.—more than a quarter of an inch per day.

Indians call the Chinook winds "snow eaters." Many Colorado winds have similar capacity and taste of the Chinook. I have seen a five-inch snowfall evaporate in a few hours from a windswept meadow without ever wetting the ground. A hundred similar instances could be given to show how moisture flies away on the wind. Snow greatly suffers from wind evaporation. In England the haymakers toss the hay about, turning it over and over, in order to get the moisture out of it. Now, wherever wind has a chance it blows snow about, shifts and re-drifts it until much of the moisture in it is evaporated and lost. But wind will suck moisture out of a solid snowdrift. I have many times tested, or assayed, the snow in drifts, in order to find the quantity of water contained in them. Invariably I found a greater percentage of water in the drifts which were best protected from the wind. What happens to a cake of ice when it is left in the wind and sun? Wind will not only take water out of washing, but it takes it out of the streams and the ground wherever it has a chance. Colorado winds are but a great dry blotter, which takes up moisture with a touch. Now, are not forests the best windbrake and shade that can be provided? The forest not only checks evaporation, but sometimes it turns tables on the wind by distilling water from it. What Colorado needs is more water, and the thieving winds should be fought as vigorously as the misguided Jayhawkers.

Heavy rainfall on a steep, barren mountain slope often washes away much soil and some bridges. Rainfall on a forested slope first fills the twig baskets, the needle rugs, and then drains off slowly, delayed by hundreds of root damlets. But in the forest much of the water goes into the porous soil and flows long distances through Nature's pure granite conduits, where it is safe from both evaporation and pollution. The litter, the humus, which collects upon an evergreen forest floor, contains humic and other acids which have a sterilizing effect upon the ground, so that a forested slope would be more likely to issue pure water than a barren one.

A forest reserve is not only a thing of beauty and profit, but it is strictly democratic. A forest reserve is a woodland from which all kinds of "land skinners" are excluded. A reserve is woodland which is sheared and not skinned. It is a place where trees are cut and planted—where seeds fall with the chips. It means more wood, mining timbers and lumber, instead of less. It would also mean more water. A reserve is not to be simply looked at. It invites the camper, prospector, miner, forester and the lumberman. It calls for a fireman, the forest ranger. A reserve means a place where one forest is harvested and another planted and protected. A forest reserve means the end of anarchy in the forest, and in a reserve the individual who departs and leaves a campfire burning, or the cowboy, sheep herder or timber thief who fires a forest, will have to pay the penalty. Forest fires have damaged the forests five times the extent of use and of all enemies. The reserve policy will reduce fire damage to the minimum.

The climate and scenery industry pays Switzerland from fifty to one hundred million dollars each year. The boosters insist that Colorado has climate and scenery as well as Switzerland. Colorado has a great future, but among all industries the tourist industry promises to be the profitable of all. But does not the tourist and all other industries chiefly or in some way greatly depend upon the forest? The tourist industry means an excellent home market. Tourists will not only bring us their gold, but their ideas. They will give us the wealth of thought from every land and clime. Tourists, like a world's fair, mean peace and prosperity, beauty and progress. A tourist hotel located in a sheep pasture will not do a very thriving business. A hotel which stands in a forest fire desert will not hold its guests very long. Deserts have little attraction for the camper or the careworn.

People are feeling the call of the wild. They want the wild, wild world beautiful. They want the temples of the gods, bits of the forest primeval, the pure and fern-fringed brooks. They like to stand "knee-deep in June," they demand the shadow of the pines, and have them they will. Above Colorado's purple forests there are alpine meadows bannered with rare blossoms, and crags and snow go up into the blue. Timberline tells stirring stories of the forest frontier. The sunny grass plots in the forest are delightful wild gardens with flowers, crags and brooks that shine in silver. There are still bits of dark forest in which one may hear the music of the pines and the songs of white cascades. Shall we have more forest?

The forest has other enemies than fire and steel. I suppose the damage done by wet snow and lightning is unpreventable. But scientific forestry, or even practical business, would prevent many damaging land and snowslides. The Swiss people have prevented many snowslides. If forest pastures are overcrowded with horses or cattle the water supply may be polluted and young trees will die or be stunted, from teeth and hoofs. If the rising generation of children are to be cared for, the rising generation of trees must be. Where trees are desired the nursery must be sparingly pastured.

The levees along the Mississippi are sometimes destroyed by a crawfish hole. The trampling of sheep on a steel, loosely soiled mountain side often causes the mountain side to be cut to pieces by erosion. Sheep crowd together, browse on young trees, trample out seedlings and feed closely. As a whole, sheep are land skinners, and to pasture them in the mountain forests is to rob the future.

Many trees are each year killed by the gnawings of porcupine. These animals gnaw off large patches of bark and frequently girdle a tree. So far as I have examined, they prefer lodge pole pine and trees of good size. Last October, on the slope of the Mount of the Holy Cross, I counted thirty-seven trees in the space of an acre that were badly gnawed, several being completely girdled.

Trees have tongues, and the forest of Colorado are full of instructive conditions and interesting stories. In all seasons and for many years I have rambled the woods. Big trees are scarce, and it is doubtful if there is a real big tree for every high peak in the state. The forest giant of Colorado, so far as I have seen, is a tree seven feet in diameter. I have often heard of trees that were ten and twelve feet in diameter, but I have not yet found these trees. It is doubtful if there is a tree in this state that is ten feet in diameter.

As a rule, the older the forest the fewer the trees upon an acre. Probably Engelmann spruce maintain more old trees per acre than any other kind. I have counted 1,900 young Lodge Pole pine upon an acre. These trees were about twenty years of age and stood about eight feet in height. Of course they stood very thick. I lost a quantity of clothing in making the count. Most of the evergreens are of slow growth. The Engelmann spruce requires ten years for an inch diameter of growth. Lodge Pole pine seem to be the slow grower, while Silver spruce is the rapid one. Trees are slow growers, and the average sawlog cut to-day began to grow about the time Washington used his hatchet. He who prowls the woods soon ceases to blame the bad sawmill man and blames the worse forestry system. But not all sawmill men are depraved. A few are heroes. Up at the source of Clear Creek is a sawmill man who most carefully defends the woods from fire. He logs like a scientific forester and a gentleman. This man is Mr. H. H. Hassel. May his tribe increase.

There were few burned spaces in the forest when the pioneers came. Indians were careful to prevent and prompt to fight forest fires. But the Indians have a tradition that there was a "big fire" over the Rockies between three and four hundred years ago. They blame the Spaniards for the fire. Leaving the tradition out, there are many evidences of a big fire about 350 years ago. Among these evidences are very old charred logs and stumps, the character and the age of much of the present forest. Let us hope that the Fire Age is over. Colorado's woodland is covered with charcoal mummies. I wonder if the clean little Japanese would be prosperous, progressive and patriotic if their land was as disfigured as our woodland? Any way, no country has ever grown poor or unpatriotic by maintaining extensive and beautiful forests.

Though trees are slow growers, as soon as an evergreen begins to grow, it begins to save water, prevent erosion, and tends to check floods. Evergreens are beautiful.

It is doubtful if any one thing could be done that would be more generally beneficial to the people of Colorado than to cause upon the cleared and burned places once more to fall the shadow of the pines.

ST. PAUL DISPATCH

MONDAY JUNE 4, 1906

VIEWS TREES AS FRIENDS

Enos A. Mills, Nature Lover,
Eloquently Pleads for Saving of Forests

APPEALS TO CLUB WOMEN

To Awaken Interest and Put a Stop
to the Wholesale Tree Slaughter

Enos A. Mills, a well known guide in the Long's Peak, Colo., district, and an enthusiastic nature student and expert in forestry, gave an address on his favorite topic, "The Forest," before the Biennial this morning.

His address breathed of the woods and their beauties. He said:

This is a beautiful world. For its charm we are chiefly indebted to the birds and the trees. Our birds are falling like withered leaves and the American forest is a picturesque remnant—a melancholy ruin of former grandeur. When bold Columbus came, America was covered with a forest green and grand. The Mississippi valley then was a sun-kissed scene in the depths of a tree-clad continent—in a forest as wide and purple as the sea.

Where once were harmonious groupings of shining lakes, winding streams, oaks and evergreens are now blackened and broken columns of templed groves—charcoal drawings of former forest glory.

THREATENED BY STEEL

Fire and steel are still insanely busy, and those marvelously picturesque wilds of lakes and pines wherein the Father of Waters finds his source, are now threatened by the sawdust blinded giant who desires to

-29-

mutilate this splendid garden for its gold.

In the Golden state the blind sawmill Sampson is pulling down the heroic columns of the grandest tree temple in the world. The Calaveras grove of big trees is being cut and dynamited for a saw mill. This immortal grove is the landmark and the heritage of the ages, and should not be desecrated. It is yet possible to save this sublime temple. A million and a half of dollars, one-third of a warship will save it. Congress is at your command. Won't you say the word?

These splendid evergreens with their historic lore and unequaled grandeur have amplitude and poetry enough to enrich the ideals of the world. As well pull down our churches, wreck the capitol, blow up the Bunker Hill monument, or almost haul down the stars and stripes as to cut these grand old trees. They are the flags of nature—the ideal emblems of the nation.

They have endured fire, flood, drouth and earth-quake, but never hauled down their evergreen banners. They have triumphed through a thousand changing seasons, watched and waved through centuries of sunlight and storm, worn monumental robes of snow flowers, stood silent in the lonely light of thousands of autumn moons, and they are still here to inspire us with their steadfastness and their splendor. Let them live on: They will bless all who make the sacred pilgrimage to see, and be a choir invisible to all who simply know that on the sublime Sierras they still grandly wave.

A few wandering, homeless tribes upon a desert is the history of every old nation that forgot its best and oldest friend the forest.

Japan and Germany, the triumphant nations at the St. Louis world's fair, have the best cared-for forests on the earth.

CONTRIBUTE TO ADVANCEMENT

The wild forests have enormously contributed to the advancement, wealth and welfare of our race. But these wild forests are almost gone, and if we do not at once become tree planters, travelers from Japan will some day soon be sketching ruins on the Western world.

A tree is the most useful plant that grows, as well as one of the most beautiful.

Trees supply food and fire, shade and shelter. They beneficially control the flow of water and the blow of wind. They make even the Storm King calm and kind. Trees paradise the earth. More than all else they enable us to have and to hold high ideals. Says Foss:

"The woods were made for the hunters of dreams,
   The streams for the fishers of song,
To those who hunt for the gunless game
   The streams and woods belong.
"There are thoughts that moan from the soul of the pine,
   And thoughts in the flower bell curled,
And thoughts that are blown with the scent of the fern
   Are as new and as old as the world."

Whitman says that all great poems and all heroic deeds were conceived in the open air. It seems that all truly great men and women have been nature lovers. May the tribe increase! How dead and desolate would be the earth if upon it did not fall the shadow of the pines. Oh how little there is in a desert to love. Exterminate the birds and the trees and you will have a realization of all the awfulness portrayed in Byron's "Darkness." Without trees the human race would be a lost child crying in the night. Plant trees and the great globe goes happily spinning "down the ringing grooves of change."

Young America sowed wild oats with the inherited forest fortune. Are we not finding with Mr. Dinkielspiel that wild oats make poor breakfast food? Mr. Dooley says it makes no difference what kind of doctor you have so long as you have a good nurse. Our tree friends have had the worst of doctors and no nurse at all. I want you each and all to be mothers to the infant tree planting industry.

WONDERLAND WITHOUT A GUN

Won't you tell your boys that the forest is a wonderland if visited without a gun? A camera is a revelator. Tell the boys that most of our trees are planted and protected by squirrels and birds. The little gray squirrel is the chief forester and nurseryman of all the western pineries. This little squirrel is being killed by the thousands, and so too are the hawks and owls who are striving day and night to prevent the rats and rabbits from killing the baby trees, and the woodpeckers, nuthatches and the dear little chickadees, who are as busy as can be trying to caterpillars, insects and borers from destroying the trees. Above all, tell the boys to put out their campfires or from it the woods may take fire and all the trees, squirrels and birds be burned.

We need picnic grounds for both body and soul. Do we not need forests to suppress floods, to maintain the springs, to give water to the irrigator, to supply lumber for the cottage, good cheer for the hearthstone, beauty to the landscapes, climate and health for everybody and scenery for all?

More than all other factors the forest prevents violent climatic changes which are so blighting and deadening to all human, plant and animal life. The farmer or the fruit grower who is not sheltered by trees is annually damaged beyond computation by high winds and by hot and cold waves. Trees prevent the wind from blowing filthy germ-laden dust into our eyes and throats. The acid of leaves sterilize the soil and give purer water. In the irrigated West the forest is absolutely necessary in order to prevent the extra dry winds from evaporating the major portion of the precious water.

Wood has hundreds of uses and most of these are the nature of necessity. The majority of the human race live in wooden houses and cook and keep warm by wood fire. Mother Nature made coal out of wood. Wood enters into the construction of vehicles on land and ships on the sea. Most of our musical instruments are of wood, so too our furniture. Would you wish to spare olives, spices, maple sugar, chocolate or coffee? Apples, oranges, grapes, dates, and walnuts are the best of foods: they enable bachelors to do without cooks, and trees supply the orange blossoms if they cannot do with out them. Do you ever use sassafras, quinine, turpentine or camphor? Old or young, sick and well, do we not need trees externally, internally and eternally?

It will pay to plant trees. One-third of the accumulated wealth of the United States has been drawn from the forest. We are now using enormous quantities of wood and for the future we will need it even in larger quantities. We will also need extensive forests. The only way to have these is to plant them. Primitive people and pioneers could depend upon wild products for a living: wild game supplied meat, wild grass forage, and wild trees fruit, fuel, shelter and lumber. Civilized people cannot depend on the scanty and unreliable wild sources but must sow if they are to reap. Civilized people must domesticate and improve the plants and animals which it needs. A complete domestication of both birds and trees is now necessary.

MUST RECEIVE ATTENTION

All trees are soon to receive the attention that the apple tree is now receiving; and all birds should before long be so tame that they would sit on our shoulders or eat from our hands. Birds and trees are the best friends we have and is anything more encouraging than to realize that they are civilizing and winning us. Scores of questions which you are in vain trying to solve would be settled by arousing interest in birds and trees. The high price of lumber is now sentencing numbers to the slums. Plant trees and the cottage with the hollyhocks by the door will increase and send forth inspired men and women.

Forests give mankind the most helpful service by storing up water on the few rainy days for the many days on which rain does not fall. Four-fifths of the rain which falls in a forest at once goes into the ground and this keeps the springs perennial and poetic. Trees also anchor the soil. Water that falls upon a barren slope for the most part rushes off in a flood and usually carries with it quantities of soil. Whenever you read of a damaging flood, remember that it must have started upon treeless soil. Continuous forest destruction on the water sheds of the of the Mississippi accounts for this river from time to time placing the high water mark higher, and the low water mark lower. The governing forest is gone.

The rainfall on the watersheds of the Mississippi is about the same as of yore, yet the flood damage along this river is growing greater, despite the fact that millions are annually spent in dredging and in levee building. The trouble is that the forests on the water sheds have been cut away and the water that falls being without forest restraint rushes off in floods. When the water-sheds of the Mississippi are reforested, and not till then, will the Father of Water flow "unvexed to the sea."

For twenty years I have heard the call of the wild echoing through the forests and mountains of the West. When only a boy I heard Nature's bugle song on Alaska's scenic shore. Alone and unarmed I have visited the silent places in the snows and flowers. Intense and happy days I have had with only bark, berries or mushrooms to eat. But what did eating matter? I felt the occult eloquence of the tongueless scenes; the world was young. Between crag and pine the wild cataract "leapt in glory." The chickadees were confiding and the pines were my friends. Many a night, alone in the forest's depths, I have had weird reveries by a campfire and felt like a wondering ancestor in the legend-making age. Life with Nature is always real, and sometimes with storm or snowslide it is in deadly earnest. I know what it is to be alone with the moon on high peaks, and I have felt the spell that holds the lonely wanderer when on a still night he feels wistful, tender touch of the summer air while the leaves whisper and listen in the moonlight, while across the magic trail falls the moon-toned etchings of the pines.

Many a night I have lain down to sleep upon forest rugs beneath the solemnity and splendor of "the wide and starry sky," and half expected that the tangled silver braid that drifts the heavens would entangle among the tree spires. Often I have been awakened by the caress of the summer rain upon my cheek or by the velvet breath of the winter snow. Often, oh how often, I have awakened just as the fires of sunrise were fringing the east and found the dear flowers bending over to watch my face; while around me the merry birds were just beginning another busy day. I cannot transfer to you the raptures that Nature has within me stirred, but throughout all and best of all I have been sustained by the friendship and the steadfastness of the heroic trees. For ages trees have been our friends, and through the Future's golden days I hope we will be theirs.

FOREST FAMINE IMPENDS

Up through the song-filled forest we have come from the low vaulted cave to the cottage and high ideals. Within the mellow lighted forest aisles still are the paths to progress and to peace. Women of the Federation: You are a multitude of intellectuals from every section of the land we love; our land is now imperilled by a forest famine; you can if you will again cover it with a forest green and grand. Be a forest friend now and you will compel the historian of the future to record that you commanded the greatest achievement of the Twentieth century—the triumph of the trees.

The wild gardens of Nature are the best kindergartens. The child who breathes the pure air that blows among trees, birds and flowers has the greatest of advantages. Children from Nature's Book and School stand highest in the examinations of life, and carry Life's richest treasures: health, individuality, sincerity and wholesome self-reliance, sincerity and wholesome self-reliance. Children touched with Nature are natural, and, like Tiny Tim, they love everybody. If baby trees and baby children grow up together it will be a profound advantage to both. Give your child the companionship of trees, and you can go down to your grave feeling that not only your child is safe, but also that Liberty, our most sacred hope and need, will go on forever.

The pathway to the Heroic Age through the forest lies.

SUPREME COURT OF OHIO
JUDICIARY BUILDING
COLUMBUS

FLORENCE E. ALLEN

January 3, 1931.

My dear Mrs. Mills:

I cannot tell you how I appreciate receiving the reprint of Mr.Mills' article upon "Trees That Twist." With its lovely pictures of the twisted trees it brought back to me the unforgettable country around Estes Park, and our very delightful visit with you last summer. I do hope that I may have that privilege again.

We went out and tried to see the mountain sheep at Specimen Mountain, but had no luck. Some time I want to ask you for further directions about the time that we should go in order to have that wonderful experience.

Looking forward to seeing you next summer,

Cordially yours,

Florence E. Allen

Mrs. Enos A. Mills,
Estes Park, Colorado.

EATONS'
RANCH
WOLF WYOMING

↑
Arrowhead

−‖
Bar Eleven

EATON BROTHERS
WOLF, WYOMING

Dec. 22,1930

Dear Mrs. Mills:

When Mr. Alden Eaton left for Arizona he told us he wanted to go over his list and send greetings to his many friends in the West but he was too tired out.   We did not press it although we knew he would if there was any insistence.

It seems to us as we look back over the losses that have come through time- Mr. Howard Eaton in 1922 and as we remember Mr. Mills that same year- Mr. Willis Eaton in 1929- these men were always so eager to do things that they went beyond their strength. They so enjoyed their work that to ease off was the only hardship.

We are sending the little booklet of the Trees That Twist down to Mr. Eaton and know he will appreciate it very much and to have your remembrance.   With many thanks and kindest wishes for your continued success

Yours truly

John Fleming

LECTURE

To=Night

AT

Carnegie Hall

February 22nd, at 8 o'clock p. m.

ON

"Our Friends, The Trees"

BY

HON. ENOS A MILLS

OF COLORADO.

This address will be on the subject of "Forestry" and by one who is an official of the United States Agricultural Department in Washington. Mr. Mills is an able speaker and master of his subject.

Admission Free

UNION LABEL Courier and Times Press.